Student Support Materials for

AQA
AS **PHYSICS**
SPECIFICATION (A)

Module 2: **Mechanics and Molecular Kinetic Theory**

Dave Kelly
Series editor: John Avison

This booklet has been designed to support the AQA (A) Physics AS specification. It contains some material which has been added in order to clarify the specification. The examination will be limited to material set out in the specification document.

Published by HarperCollins*Publishers* Limited
77–85 Fulham Palace Road
Hammersmith
London
W6 8JB

> www.**Collins**Education.com
> Online support for schools and colleges

© HarperCollins*Publishers* Limited 2000
First published 2000
Reprinted 2000, 2001

ISBN 0 00 327716 X

Dave Kelly asserts the moral right to be identified as the author of this work

All rights reserved. No part of this publication may be reproduced, stored in a retrieval system, or transmitted in any form or by any means, electronic, mechanical, photocopying, recording or otherwise, without either the prior permission of the Publisher or a licence permitting restricted copying in the United Kingdom issued by the Copyright Licensing Agency Ltd., 90 Tottenham Court Road, London W1P 0LP.

British Library Cataloguing in Publication Data
A catalogue record for this publication is available from the British Library

Cover designed by Chi Leung
Typesetting by Derek Lee
Printed and bound by Scotprint, Haddington

The publisher wishes to thank the Assessment and Qualifications Alliance for permission to reproduce the examination questions.

You might also like to visit

> www.**fire**and**water**.com
> The book lover's website

> *Other useful texts*
>
> **Full colour textbooks**
> *Collins Advanced Modular Sciences: Physics AS*
> *Collins Advanced Science: Physics*
>
> **Student Support Booklets**
> *AQA (A) Physics: 1 Particles, Radiation and Quantum Phenomena*
> *AQA (A) Physics: 3 Current Electricity and Elastic Properties of Solids*

To the student

What books do I need to study this course?

You will probably use a range of resources during your course. Some will be produced by the centre where you are studying, some by a commercial publisher and others may be borrowed from libraries or study centres. Different resources have different uses – but remember, owning a book is not enough – it must be *used*.

What does this booklet cover?

This *Student Support Booklet* covers the content you need to know and understand to pass the module test for AQA (A) Physics Module 2: Mechanics and Molecular Kinetic Theory. It is very concise and you will need to study it carefully to make sure you can remember all of the material.

How can I remember all this material?

Reading the booklet is an essential first step – but reading by itself is not a good way to get stuff into your memory. If you have bought the booklet and can write on it, you could try the following techniques to help you to memorise the material:

- underline or highlight the most important words in every paragraph
- underline or highlight scientific jargon – write a note of the meaning in the margin if you are unsure
- remember the number of items in a list – then you can tell if you have forgotten one when you try to remember it later
- tick sections when you are sure you know them – and then concentrate on the sections you do not yet know.

How can I check my progress?

The module test at the end is a useful check on your progress – you may want to wait until you have nearly completed the module and use it as a mock exam or try questions one by one as you progress. The answers show you how much you need to do to get the marks.

What if I get stuck?

A colour textbook such as *Collins Advanced Modular Sciences: Physics AS* provides more explanation than this booklet. It may help you to make progress if you get stuck.

Any other good advice?

- You will not learn well if you are tired or stressed. Set aside time for work (and play!) and try to stick to it.
- Don't leave everything until the last minute – whatever your friends may tell you it doesn't work.
- You are most effective if you work hard for shorter periods of time and then take a (short!) break. 30 minutes of work followed by a five or ten minute break is a useful pattern. Then get back to work.
- Some people work better in the morning, some in the evening. Find out which works better for you and do that whenever possible.
- Do not suffer in silence – ask friends and your teacher for help.
- Stay calm, enjoy it and … good luck!

There are rigorous definitions of the main terms used in your examination – memorise these exactly.

The examiner's notes are always useful – make sure you read them because they will help with your module test.

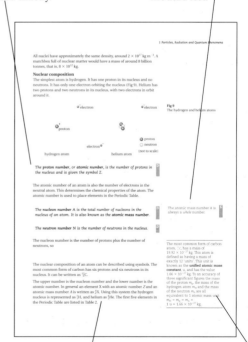

The main text gives a very concise explanation of the ideas in your course. You must study all of it – none is spare or not needed.

Further explanation references give a little extra detail, or direct you to other texts if you need more help or need to read around a topic.

11.1 Mechanics

11.1.1 Scalars and vectors

Physical quantities can be classified into two groups: **scalars** or **vectors**. Scalar quantities, such as temperature or mass, have a magnitude (size) but have no direction associated with them. Scalars can be fully described by a single number and a unit. For example, stating that the room temperature is 20 °C or that the mass of a person is 80 kg fully specifies these quantities. Other physical quantities, like velocity or force, have a direction associated with them. These are known as vector quantities. A vector quantity is only fully specified when the magnitude *and* the direction are given. It isn't sufficient to know that a force has a magnitude of 300 N; we also need to state what direction it acts in, e.g. a force of 300 N acting horizontally in a direction 30° east of north.

> **E** If you are asked to find an unknown force or velocity, don't forget to give the direction as well as the magnitude.

Table 1 Examples of scalar and vector quantities met in this module

Scalars	Vectors
distance	displacement
speed	velocity
energy	force
power	acceleration
temperature	momentum
mass	

> **E** Make sure that you know which quantities are vectors and which are scalars.

> **D** *A vector quantity has magnitude and direction, whereas a scalar quantity has magnitude only.*

Vector quantities are often identified by the use of **bold type**, such as *s* for displacement.

Adding vector quantities

When two vectors are added, we need to take account of their direction as well as their magnitude. Two vectors can be added by drawing a scale diagram showing the effect of one vector followed by the other, i.e. by drawing them 'nose to tail' (see Fig 1).

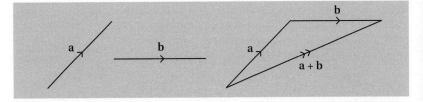

Fig 1 Adding vectors
The sum of **a** and **b** is found by drawing a and b so that the arrows showing their direction follow on

The sum of a number of vectors is known as the **resultant**. The resultant is the single vector that has the same effect as the combination of the other vectors. It is vital to take into account the relative direction of vectors when adding them together, for example the resultant of two 5 N forces could be anything from zero to 10 N, depending on their directions.

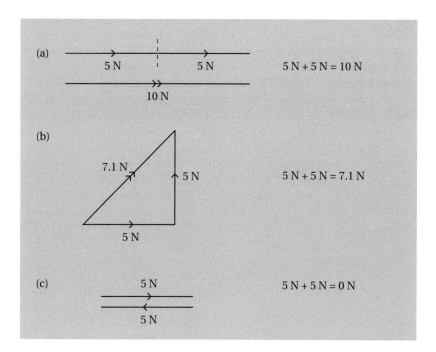

Fig 2
The magnitude and direction of the resultant depends on the orientation of the two component vectors

The resultant of two vectors can also be found by the **parallelogram law**. A parallelogram is constructed using the two vectors as adjacent sides. The resultant is the diagonal of the parallelogram (Fig 3).

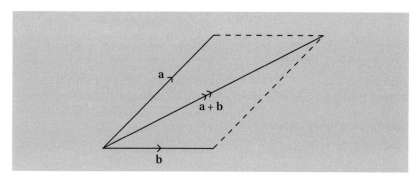

Fig 3
The parallelogram method for finding the resultant

If the vector diagram is drawn to scale, the resultant vector can be found by direct measurement from the diagram.

For two vectors at right angles, the magnitude of the resultant can also be found from calculation using Pythagoras' theorem (Fig 4).

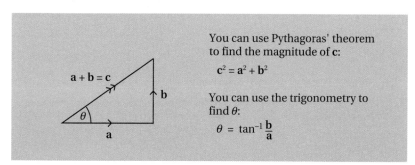

You can use Pythagoras' theorem to find the magnitude of **c**:

$$c^2 = a^2 + b^2$$

You can use the trigonometry to find θ:

$$\theta = \tan^{-1}\frac{b}{a}$$

Fig 4
Adding vectors at right angles

The angle of the resultant for vectors at any angle to each other can be found using the sine rule and the cosine rule, but these methods are beyond the AS specification.

Example

A tanker is being pulled into harbour by a tug boat which exerts a force of 200 MN in an easterly direction. The tanker is also subject to a force 150 MN due to a northerly current. Find the resultant force acting on the tanker.

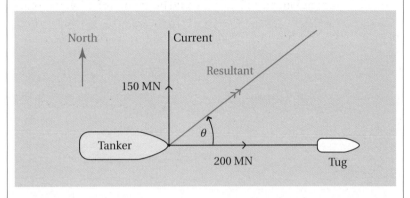

Answer

The magnitude of the resultant, **R**, is given by the equation:

$R^2 = 200^2 + 150^2$

$= 40\,000 + 22\,500$

$= 62\,500$

R = 250 MN

Find the direction of the resultant by:

$\theta = \tan^{-1} \dfrac{150}{200}$

$= \tan^{-1} 0.75$

$= 37°$ north of east

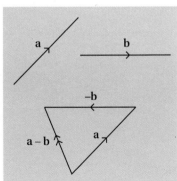

Fig 5
Subtracting a vector

Subtracting a vector quantity can be thought of as adding a negative vector. The vector which is to be subtracted is reversed in direction. This reversed, or negative, vector is then added in the usual way (Fig 5).

Example

A river is flowing at 1 m s^{-1}. Find the speed and direction that a swimmer must travel in if they are to achieve a resultant velocity of 1.5 m s^{-1} directly across the river.

Answer

The swimmer's velocity, **v**, is the resultant minus the river's velocity.

Magnitude of velocity = $\sqrt{1.5^2 + 1^2}$

$= 1.8$ m s^{-1}

Direction of velocity, $\theta = \tan^{-1} \dfrac{1}{1.5}$

$= 34°$

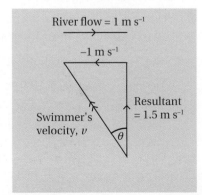

Resolution of vectors

A single vector can be replaced by a combination of two or more vectors that would have the same effect. This process is called **resolving** the vector into its **components** and it can be thought of as the reverse of finding the resultant. The components of a vector could be at any angle but it is often useful to use two components that are at right angles to each other. This might be to find the horizontal and vertical components of a force or a velocity (Fig 6).

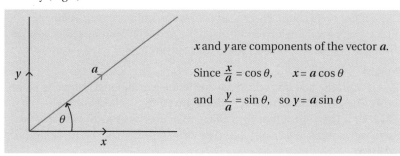

x and y are components of the vector a.

Since $\frac{x}{a} = \cos\theta$, $x = a\cos\theta$

and $\frac{y}{a} = \sin\theta$, so $y = a\sin\theta$

Fig 6
Calculating perpendicular components

Example 1

A wind is blowing at 15 m s⁻¹ in a north-easterly direction. Find the components of the wind's velocity which blow towards the north and towards the east.

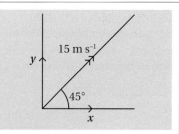

Answer:

The northerly component, $y = 15\sin 45° = 10.6$ m s^{-1}
The easterly component, $x = 15\cos 45° = 10.6$ m s^{-1}

Example 2

A car of weight 10 000 N is parked on a steep hill which makes an angle of 20° to the horizontal. Resolve the car's weight into components that act along the slope and at 90° to the slope.

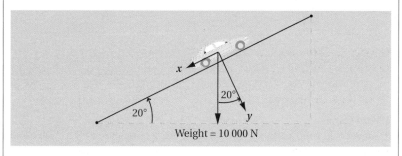

Weight = 10 000 N

Down the slope, component, $x = 10\,000\sin 20° = 3420$ N
Along the slope, component, $y = 10\,000\cos 20° = 9497$ N

> **E** Remember that the vector you are resolving is always the hypotenuse of a triangle. The components will always be smaller than the original vector.

11.1.2 Conditions for equilibrium with two or three coplanar forces acting at a point

It is often important to be able to identify, and add together, the forces that are acting on an object. The size and direction of the resultant force will determine what happens to the object.

Forces

Everyday objects are subjected to a variety of forces, such as weight, contact forces, friction, tension, air resistance and buoyancy. All these forces, except weight, are electromagnetic in origin. They arise because of the attraction or repulsion of the charges in atoms.

Weight

This is the force that acts on a mass due to the gravitational attraction of the Earth. The gravitational field strength, g, on Earth is approximately $9.81 \, \text{N kg}^{-1}$. This means that each kilogram of mass is attracted towards the Earth with a force of 9.81 newtons. The weight of an object (in newtons) is given by:

weight (N) = mass (kg) × gravitational field strength (N kg^{-1}) *or* $W = mg$

The total weight of a real object is the sum of the gravitational attractions acting on every particle in the object. The resultant of all these forces is the weight of the object which can be treated as a single force acting at a single point in the object. This point is called the **centre of gravity.**

> **E** Gravitational field strength varies slightly from place to place on the Earth's surface, and decreases as you move away from the Earth.
>
> Gravitational field strength on the moon is about $1.6 \, \text{N kg}^{-1}$, about one sixth of its value on Earth. On the moon, you would weigh one sixth of your weight on Earth, though your mass would remain the same.
>
> Gravitational field strength, measured in N kg^{-1}, is represented by the letter g. The symbol g is also used to represent the acceleration due to gravity and can be given in m s^{-2}. These quantities are numerically equivalent.

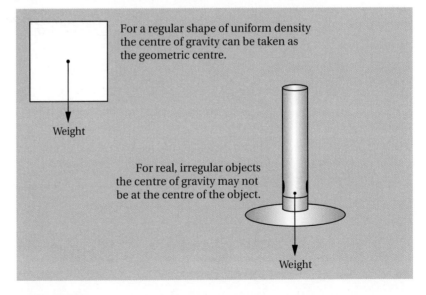

Fig 7
Centre of gravity
For a regular shape, like a sphere or a cube, the centre of gravity is taken as the geometric centre. The centre of gravity in irregular objects depends on the arrangement and densities of materials in the object

Contact forces

Whenever two solid surfaces touch, they exert a contact force on each other. This force is often known as the **reaction.** It is the contact force between the floor and your feet that stops gravity pulling you through the floor. The resultant contact force between two surfaces could be at any angle (Fig 8).

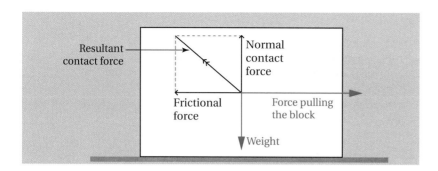

Fig 8
Forces on an object moving across a surface

We usually split the force into two components:
- the normal contact force acting perpendicularly to the two surfaces
- the frictional force, acting parallel to the surfaces.

Friction
A frictional force acts between two surfaces whenever there is relative motion between them, or when an external force is trying to slide them past each other.

Tension
An object is said to be in tension when a force is acting to stretch the object. Elastic materials, like ropes or metal cables, resist this stretching and exert a force on the bodies trying to stretch them.

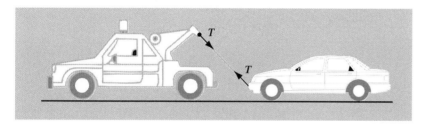

Fig 9
The tension in the tow-rope, T, acts to pull the truck backwards and to pull the car forwards

Air resistance
Any object that is moving through a fluid is subject to a resistive force or **drag**. Any object moving through the atmosphere has to push the air out of the way; this gives rise to the drag force that acts to oppose relative motion between the object and the fluid.

Buoyancy
Any objects that are partly or fully submerged in a fluid, like a boat floating on water or a hot-air balloon floating in the atmosphere, are subject to an upthrust from the surrounding fluid.

The size of the air resistance acting on an object depends on the area of the object and on the density of the air. The air resistance increases approximately as the *square* of the relative speed between the object and the air. So, as you go faster, the force trying to stop you increases much more rapidly than your speed.

Free body diagrams
The forces acting on a real object may be quite complex. A **free body** diagram is an attempt to model the situation so that we can analyse the effect of the forces. The free body diagram is used to show all the external forces that are acting on an object. Since forces are vector quantities they are represented by arrows, drawn to scale and acting in the right direction.

Example

Draw a free body diagram for:

(i) A hot air balloon tethered by a cable.

Answer

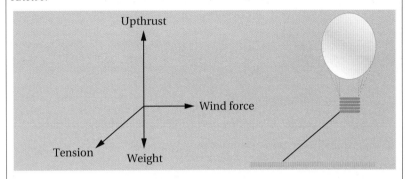

> **E** Make sure that you only include forces acting on the object you are considering, e.g. the child on the slide. Don't complicate things by including forces that act on other objects, e.g. the slide.
>
> Don't complicate things by including internal forces, like the tension in the child's muscles.

In this specification, problems will be restricted to systems of two or three **coplanar** forces. Coplanar forces act in the same plane. In other words you only need to consider two-dimensional problems.

Example

(ii) A child sliding down a playground slide

Answer

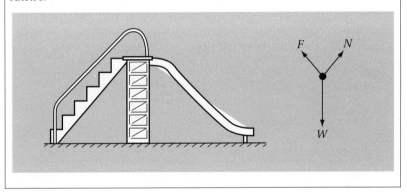

Fig 10
a + **b** + **c** must form a closed triangle if the body is to be in equilibrium

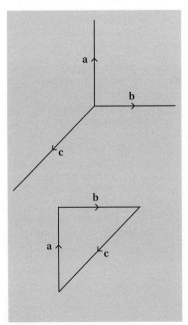

Equilibrium

Once we have identified all the forces acting on an object and drawn a free body diagram, we can use vector addition to find the resultant force. If the resultant force is not zero then the object will accelerate in the direction of the resultant force. If an object is not accelerating, it is said to be in **equilibrium**.

> **D** *An object is said to be in equilibrium if it is stationary, or moving at constant velocity.*

This idea is expressed in Newton's first law of motion (see page 25). One of the conditions for equilibrium is that all the external forces that act upon the object must add up to zero. This means that if three forces are acting, the vector addition must form a closed triangle (Fig 10).

Example

A tightrope walker of mass 50 kg is standing on the middle of a tightrope. The rope makes an angle of 15° with the horizontal. Draw a free body diagram for the tightrope and find the tension in the rope. (Assume that the mass of the rope is negligible and take $g = 10\ N\,kg^{-1}$)

Answer

Since the contact force between the rope and the person must balance the person's weight, the contact force, F, acting on the rope is 500 N. If the rope is to stay in equilibrium, the vector diagram must form a closed triangle.

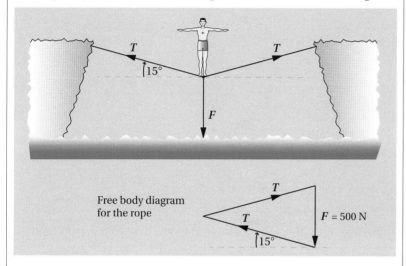

Free body diagram for the rope

Since the situation is symmetrical the tension, T, is given by:

$$\frac{250}{T} = \sin 15°$$

$$T = \frac{250}{\sin 15°} = 966\ N$$

An alternative method of investigating these problems is to resolve all the forces into two perpendicular directions, often horizontal and vertical. If the object is to be in equilibrium, two conditions must be satisfied:

- the sum of the horizontal components must be zero;
- the sum of the vertical components must be zero.

Example

A car of mass 1200 kg is parked on a hill inclined at 20° to the horizontal. The maximum frictional force between the tyres and the road is 5000 N. Will the car remain in equilibrium?

Answer

First we need to resolve W into components that are perpendicular and parallel to the slope. For equilibrium perpendicular to the slope:

$R = W \cos 20° = 12\,000 \times 0.94 = 11\,300\,\text{N}$

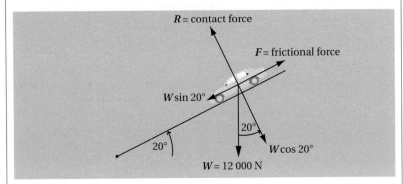

For equilibrium parallel to the slope:

$F = W \sin 20° = 4100\,\text{N}$

This is less than the maximum value; the car will remain in equilibrium.

11.1.3 *Turning effects*

Moments

Forces can cause objects to accelerate in a straight line. However, forces can also have the effect of turning or tipping an object. Even when an object is acted upon by two equal and opposite forces, it may not be in equilibrium. If the forces do not pass through a single point, the object will tend to rotate. For example if you push a wardrobe at the top, and there is a large frictional force due to the carpet at the bottom, the wardrobe will tip rather than slide (Fig 11).

Fig 11
The moment about X causes the wardrobe to tip over

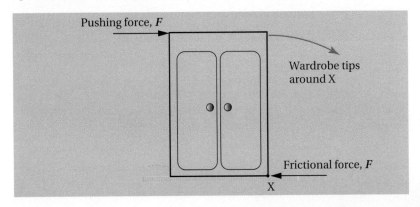

The turning effect of a force about a point is known as its **moment** or **torque**. The moment of a force about a point depends on two things:

- the magnitude of the force;
- the perpendicular distance from the line of the force to the point.

> The moment of a force about a point is equal to the magnitude of the force, F, multiplied by the perpendicular distance of the force from the pivot, s.
>
> moment (N m) = s (m) × F (N)

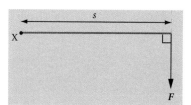

Fig 12
The moment of force F about X is given by
moment = sF

The moment of a force is measured in newton metres.

You can increase the torque of a spanner on a nut by exerting a larger force, or by getting a longer spanner. When cycling a bicycle, the maximum turning effect is when the pedal crank is horizontal. When the pedal crank is vertical there is no moment, since the force passes through the pivot.

It is a common mistake to give the units for the moment of a force as Newtons *per* metre (N/m). This is incorrect, the right units are Nm.

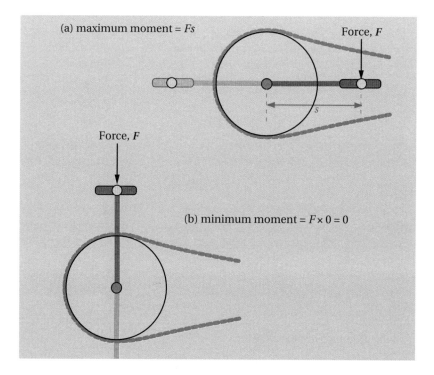

Fig 13
Moments in a pedal system

It is important to note that it is the **perpendicular** distance from the pivot to the line of the force. In Fig 14, where a vertical force is being used to lift a trap door, the moment is $Fs \cos \theta$

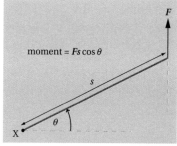

Fig 14
The moment required to open the trap door at any instant is $Fs \cos \theta$

The principle of moments

There is a second condition that must be satisfied if an object is to be in equilibrium under the action of several forces. Not only must the sum of the forces be zero, the sum of the moments about any point must also be zero. If this condition is not met, the object would rotate around that point.

> If an object is in equilibrium the sum of the moments about any point must be zero.

Another way of putting this is to say that the sum of the moments which tend to turn the object anticlockwise must be equal to the sum of the clockwise moments. The most straightforward example of this is on a see-saw. A heavier child can be balanced by a lighter child if the lighter child sits further from the pivot (Fig 15).

Fig 15
Equilibrium on a see-saw

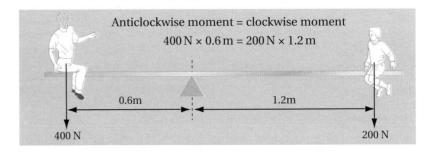

The principle of moments can be applied to find the magnitude of unknown forces.

> *Example*
>
> *A crowbar (lever) is used to lift a paving slab which weighs 300 N. The crowbar pivots at a point 0.20 m from the slab. How much force will it take to lift the slab, if the force is applied 1.2 m away from the pivot?*
>
>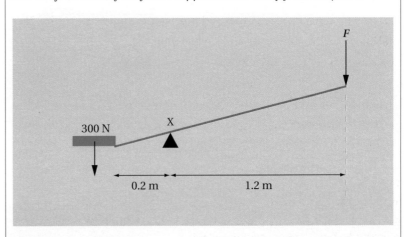
>
> *Answer*
>
> Taking moments about the pivot, X:
>
> Anti-clockwise moment = 300 N × 0.2 m = 60 N m
>
> Clockwise moment = F × 1.2 m = 1.2 F N m
>
> If the crowbar is in equilibrium these moments must balance:
>
> 60 = 1.2 F, so $F = \dfrac{60}{1.2} = 50$ N
>
> To lift the slab, the force must be just greater than 50 N.

E This lever gives you the ability to lift large weights with a smaller force. This is known as a **mechanical advantage**. However, you will have to move your force much further than the slab will move. The **work** that you do will never be less than the **energy** gained by the slab.

Sometimes it is necessary to apply *both* the conditions for equilibrium in order to calculate all the forces in a situation:

- The vector sum of the forces must be zero
- The sum of the moments about any point must be zero.

> *Example*
>
> *Two people are carrying a 3 m long plank which has a mass of 20 kg. Graham is holding the plank at one end, whilst Beryl is holding the plank 1 m from the opposite end. Calculate the forces which each must exert. (Take $g = 10\,N\,kg^{-1}$)*
>
>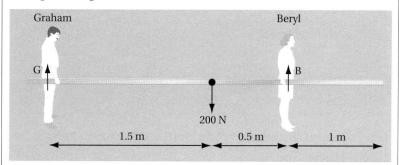
>
> *Answer*
>
> For equilibrium, the sum of the vertical forces are equal: $G + B = 200\,N$
>
> Taking moments about Graham's end of the plank:
>
> Clockwise $200 \times 1.5 = 300\,N\,m$
>
> Anticlockwise $B \times 2 = 2B\,N\,m$
>
> For equilibrium these must be equal: $2B = 300\,N\,m$
>
> So Beryl's force is 150 N. The rest of the 200 N plank is supported by Graham, a force of 50 N.

Always draw a diagram showing the forces acting.

Take care when choosing a point to take moments about. Choose a point that one of the unknown forces passes through so that this force has no moment about that point.

Couples

Two parallel forces which act in opposite directions will tend to make an object rotate without moving its position. If these forces are equal in magnitude, they are known as a **couple**.

> The turning effect, or torque, of a couple is Fs.

A common example of a couple is in an electric motor where the force on each side of the coil is equal but opposite (Fig 17).

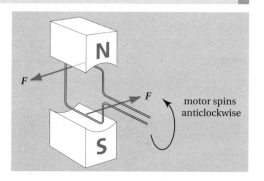

Fig 17
A couple in an electric motor

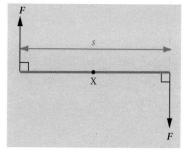

Fig 16
A couple
The torque of a couple about the midpoint is

$$F \times \left(\frac{s}{2}\right) + F \times \left(\frac{s}{2}\right) = Fs$$

E In a place where gravity is uniform, the centre of mass and the centre of gravity are in the same place.

Centre of mass

All the mass of a body can be thought of as acting at a single point known as the centre of mass of a body. If the resultant force on an object passes through the centre of mass it will accelerate without rotating. If the resultant force does not pass through the centre of mass the object will spin (Fig 18).

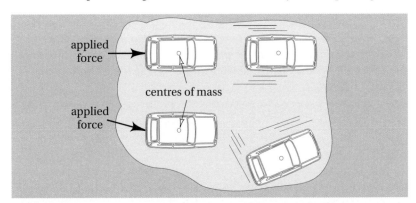

Fig 18
Centre of mass
The applied force in the lower car does not pass through the centre of mass so the car spins

11.1.4 Displacement, speed, velocity and acceleration

Displacement and distance

Displacement and distance are both measured in the same units, metres, but displacement, *s*, is a vector quantity that describes the *effect* of a journey rather than the total distance travelled. Distance travelled is a scalar quantity.

Fig 19
The vector *s* represents the displacement. This is the net effect of the journey, and in this case has a much smaller magnitude than the distance travelled.

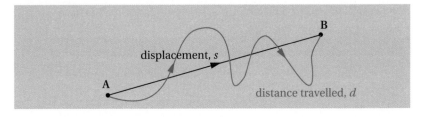

D Displacement, s, is the distance travelled in a given direction.

Speed and velocity

Speed is the distance covered in a certain time.

$$\text{speed (m s}^{-1}\text{)} = \frac{\text{distance travelled (m)}}{\text{time taken (s)}}$$

Speed is a scalar quantity which is measured in metres per second, or kilometres per hour. During a journey the speed may be changing. The average speed over the whole journey is given by:

$$\text{average speed} = \frac{\text{total distance covered}}{\text{total time taken}}$$

E The greek letter Δ represents a change in a physical quantity; Δs is a small change in the displacement of an object.

The speed at any given instant in the journey may be above or below the average speed. The speed at a certain time is known as the **instantaneous speed**. If we measure the distance covered in a very small time interval, Δt, the value for speed approaches the instantaneous value.

Velocity is a vector quantity, it has a magnitude (measured in m s^{-1} or km h^{-1}) *and* a direction. Velocity is the speed in a given direction and is defined by:

$$\text{velocity (m s}^{-1}) = \frac{\text{displacement (m)}}{\text{time (s)}} \qquad v = \frac{\Delta s}{\Delta t}$$

Acceleration

Acceleration is the rate at which velocity changes.

$$\text{acceleration} = \frac{\text{change in velocity}}{\text{time taken for change}} = \frac{\Delta v}{\Delta t}$$

Since the change in velocity is measured in m s^{-1}, and time is measured in seconds, acceleration is measured in m s^{-2}. Acceleration is a vector quantity and therefore takes place in a particular direction. Any change in velocity, either speeding up, slowing down or simply changing direction, is an acceleration.

Acceleration does not always take place in the same direction as the velocity. A ball thrown in the air which rises and then falls again is always accelerating downwards due to gravity (Fig 20).

> *Example*
>
> *A car goes from rest to 60 mph in 4.7 seconds. Calculate its acceleration. (1 mile = 1.6 km)*
>
> *Answer*
>
> The car's final velocity is $\dfrac{(60 \times 1600)}{(60 \times 60)} = 27 \text{ m s}^{-1}$
>
> acceleration $= \dfrac{\Delta v}{\Delta t} = \dfrac{27}{4.7} = 5.7 \text{ m s}^{-2}$

> A car driving round a roundabout may be travelling at a steady speed but it is constantly changing its velocity, because it is changing its direction.

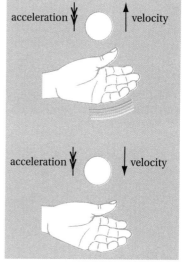

Fig 20
Acceleration and velocity of a ball thrown upwards (top) and falling back (bottom)

11.1.5 Uniform and non-uniform acceleration: Graphical methods

Displacement–time graphs

A journey can be represented by a graph showing displacement against time. The gradient of the graph represents the displacement in a certain time, which is the velocity. A straight line represents constant velocity.

 The gradient of a displacement–time graph is the velocity

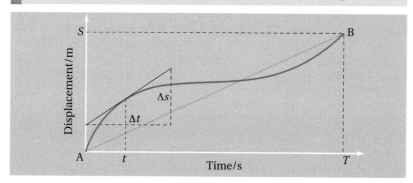

Fig 21
Instantaneous velocity and average velocity.
(i) The instantaneous velocity at time *t* is the gradient of the curve at that point, $\Delta s/\Delta t$
(ii) The average velocity for the whole journey is the gradient of the straight line drawn from A to B,
$\dfrac{S}{T}$

AQA (A) Physics AS

Example

The displacement–time graph below shows the motion of a car over 10 seconds. Describe what is happening at each stage of the journey.

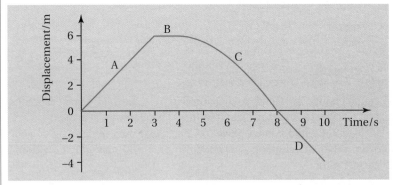

A Car moves with a constant velocity of $\frac{6}{3} = 2\,\text{m s}^{-1}$

B Between 3 and 4 seconds, the car is stationary.

C The car returns to its original position with non-uniform velocity. The car moves slowly at first and then more quickly.

D The car moves in the opposite direction (negative displacement) with a uniform velocity of
$\frac{4}{2} = 2\,\text{m s}^{-1}$

Velocity–time graphs

A velocity–time graph for a journey can be used to calculate the acceleration and the displacement. The gradient of the graph is $\frac{\Delta v}{\Delta t}$, which is acceleration. A straight line represents constant acceleration. A line with a negative gradient represents a negative acceleration. This could be slowing down (deceleration or retardation) or it could mean that the object is speeding up in the opposite direction. The area below the graph gives the total displacement.

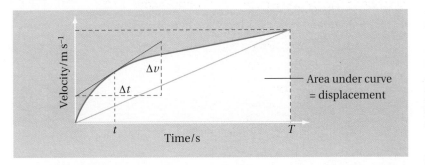

Fig 22
Velocity–time graph, showing acceleration and displacement. The acceleration at time t (gradient = $\frac{\Delta v}{\Delta t}$) is greater than the average acceleration over the time period T.

Example:

The velocity–time graph below shows a person's journey. Describe each section of the journey as fully as possible. Calculate the displacement during the first 7 seconds.

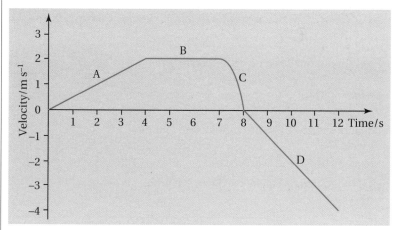

A Uniform acceleration of $\frac{2}{4} = 0.5 \text{ m s}^{-2}$

B Constant velocity of 2 m s^{-1}

C Slowing down (negative acceleration) but at a varying rate

D Negative acceleration, the person is now speeding up in the opposite direction.
   Acceleration $= -\frac{4}{4} = -1 \text{ m s}^{-1}$

Displacement is the area under the graph $= \left(\frac{1}{2} \times 4 \times 2\right) + (3 \times 2) = 10 \text{ m}$

Equations of uniformly accelerated motion

We will just consider objects that moves in a straight line with uniform acceleration. The five important variables that are used to describe this motion are shown in the table.

There are a number of equations which link these variables together and describe uniformly accelerated, straight-line motion.

Quantity	Symbol
displacement	s
initial velocity	u
final velocity	v
acceleration	a
time	t

1 The definition of acceleration is:

   $\text{acceleration} = \frac{\text{change in velocity}}{\text{time}}$, or $a = \frac{(v-u)}{t}$

   Rearranging this gives:

   $v = u + at$

2 The definition of average velocity is:

   $\text{average velocity} = \frac{\text{displacement}}{\text{time}}$

But if the velocity changes at a constant rate we can say that average velocity is:

$$\frac{(v+u)}{2}$$

So $\frac{(v+u)}{2} = \frac{s}{t}$ or $s = \frac{1}{2}(u+v)t$

> **E** Some people refer to these equations as the 'suvat' equations. It is a good idea to start each problem by writing down 'suvat'. Then you should identify which of the variables you have values for, and which equation you need to use.

3 Equations 1 and 2 can be combined to give:

$$s = ut + \tfrac{1}{2}at^2$$

4 Equations 3 and 1 can be combined to eliminate t:

$$v^2 = u^2 + 2as$$

These four equations can be used to solve problems about motion.

Example

A sprinter accelerates from rest to 11 m s⁻¹ in the first 4 seconds of a race. Assuming that his acceleration is constant, find the acceleration and the distance covered in the first 4 seconds.

Answer

$s = ?$ $u = 0 \,\text{m s}^{-1}$ $v = 11 \,\text{m s}^{-1}$ $a = ?$ $t = 4 \,\text{s}$

To find acceleration we can use the equation:

$v = u + at;\ a = \dfrac{v-u}{t} = \dfrac{11}{4} = 2.75 \,\text{m s}^{-2}$

To find the displacement we can use $s = ut + \tfrac{1}{2}at^2$

$s = 0 \times 4 + \tfrac{1}{2} \times 2.75 \times 4^2 = 22 \,\text{m}$

Acceleration due to gravity

> **E** The acceleration due to gravity varies from place to place; it is greater at the poles than at the equator.

A falling object accelerates towards the Earth because of the Earth's gravity. The acceleration due to gravity is independent of the object's mass. Experiments show that on Earth all objects, regardless of mass, accelerate under gravity at around $9.81 \,\text{m s}^{-2}$.

A falling object, dropped from an aircraft, for example, soon reaches a high velocity. After 1 second it will be travelling at $9.8 \,\text{m s}^{-1}$, and after 10 seconds it will have reached $98 \,\text{m s}^{-1}$. If no other forces were acting the object would keep accelerating until it hit the Earth's surface. However, on Earth there is always some air resistance which opposes the motion of the falling object. Air resistance, or drag, has little effect on compact objects moving at low speeds, but the drag force increases with speed and eventually it is equal to the gravitational force pulling the object towards Earth. When this happens there will be no more acceleration and the object will continue to fall at a constant velocity, known as the **terminal velocity** (Fig 23).

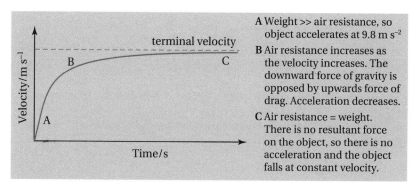

Fig 23
Velocity–time graph for an object falling in the Earth's atmosphere

A Weight >> air resistance, so object accelerates at 9.8 m s^{-2}
B Air resistance increases as the velocity increases. The downward force of gravity is opposed by upwards force of drag. Acceleration decreases.
C Air resistance = weight. There is no resultant force on the object, so there is no acceleration and the object falls at constant velocity.

> Cars have a top speed because air resistance and other resistive forces increase with speed. When the drag force is equal to the maximum driving force that a car can produce, the car will no longer accelerate. It has reached its top speed.

Terminal velocity depends on the area of the object, as well as its mass, and on the density of the air. A falling feather floats down at less than 1 m s^{-1} whilst a free fall parachutist may reach a terminal velocity of 65 m s^{-1} (close to 50 mph).

11.1.6 Independence of vertical and horizontal motion

An object fired through the air follows a parabolic path. Even though this is not a straight line, we can still use the equations of motion. This is because horizontal motion does not affect vertical motion.

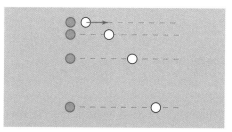

Fig 24
An object dropped vertically and one thrown horizontally will fall at the same rate

A two dimensional problem can be solved by treating it as two one-dimensional problems, i.e. keeping the horizontal and vertical motions separate. If an object is thrown with initial speed 20 m s^{-1} at an angle of 45° to the horizontal, we can calculate its range and the maximum height it will reach.

> A bullet is fired horizontally from a rifle at the same time as a bullet is dropped vertically from the same height. The two bullets will hit the ground at the same time.

The problem can be split into two parts, the horizontal motion and the vertical motion. The initial velocity can be resolved into horizontal and vertical components (Fig 25).

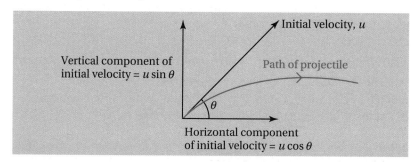

Fig 25
Components of velocity for a projectile

> Because displacement, velocity and acceleration are all vector quantities it is important to decide on a sign convention. When you start a problem you need to decide whether up is positive or negative and then stick to this throughout the problem.

If air resistance can be ignored, there is no horizontal acceleration, so the initial and final horizontal velocities are the same.

Horizontal motion	Vertical motion
s	s
$u = 20 \cos 45° \, \text{m s}^{-1} = 14.1 \, \text{m s}^{-1}$	$u = 20 \sin 45° = 14.1 \, \text{m s}^{-1}$
$v = 20 \cos 45° \, \text{m s}^{-1} = 14.1 \, \text{m s}^{-1}$	$v =$
$a = 0 \, \text{m s}^{-2}$	$a = -9.8 \, \text{m s}^{-2}$ (down is negative)
t	t

Consider the vertical velocity first. When the object reaches its greatest height its velocity will be zero: $v = 0 \, \text{m s}^{-1}$

Using $v = u + at$, or $t = \dfrac{(v - u)}{a}$

$t = \dfrac{-14.1}{-9.8} = 1.44 \, \text{s}$

We can find **s** at that time using $v^2 = u^2 + 2as$

$s = \dfrac{(v^2 - u^2)}{2a} = \dfrac{-199}{-19.6} = 10.2 \, \text{m}$

Then consider the horizontal part. The time of flight is twice the time taken to reach the greatest height, so $t = 2 \times 1.44 = 2.88 \, \text{s}$.

$s = ut + \tfrac{1}{2}at^2 = 14.1 \times 2.88 = 40.6 \, \text{m}$

Momentum

Momentum can also be given in units of newton seconds (N s); this is exactly equivalent to kg m s^{-1}.

The momentum of a moving object is defined as its mass multiplied by its velocity. Momentum is a vector quantity, so its direction is the same as its velocity. There is no special name for the unit of momentum; its units are those of mass × velocity, kg m s^{-1}.

> Momentum (kg m s^{-1}) = mass (kg) × velocity (m s^{-1}): or $p = mv$

The momentum of a body is a measure of how difficult it is to stop it. A heavy lorry could have a mass of 40 tonnes. When it is travelling on the motorway at $25 \, \text{m s}^{-1}$, the momentum is:

$p = 40 \times 10^3 \, \text{kg} \times 25 \, \text{m s}^{-1} = 1.0 \times 10^6 \, \text{kg m s}^{-1}$

Compare this to the momentum of a person running at top speed:

$p = 80 \, \text{kg} \times 10 \, \text{m s}^{-1} = 800 \, \text{kg m s}^{-1}$

11.1.7 Conservation of linear momentum

When there are no external forces acting on an object its linear momentum does not change. The **conservation of linear momentum** is an important principle in physics and can be stated as:

> The total linear momentum of a system is constant provided that there is no external resultant force acting.

It isn't immediately obvious that this is true. A car standing at traffic lights has no linear momentum, yet a few seconds later it has gained momentum as it pulls away from the lights. This does not contravene the conservation of momentum, since if we are just thinking about the car, the friction of the road on the car is an external force and conservation of momentum does not apply. If we consider the car and the Earth as the 'system' there are no external forces acting, so conservation of momentum must apply. As the car gains momentum in a forward direction, the Earth gains momentum in the opposite direction.

> **E** Whenever you jump into the air, you are moving the Earth a little. The Earth has to acquire a momentum that is equal but opposite to your upwards momentum. Since your mass is rather less than the Earth's, the Earth's velocity will be much smaller than yours.

Collisions

The conservation of momentum is often applied to situations where two objects collide. If we can ignore external forces, such as friction, the total momentum before the collision must be equal to the total momentum afterwards.

Suppose car A has a mass of 1500 kg and is moving at 20 m s^{-1} when it collides into car B which has a mass of 1000 kg and is moving in the opposite direction at 10 m s^{-1}. After the collision the two cars stick together. We can use the conservation of momentum to calculate the velocity after the collision.

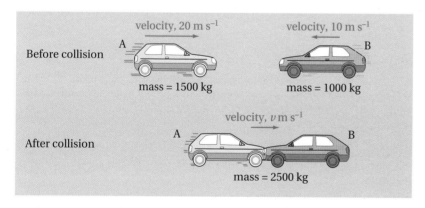

Fig 26
Conservation of momentum in a car collision

Momentum of car A = 1500 kg × 20 m s^{-1} = 30 000 kg m s^{-1}
Momentum of car B = 1000 kg × (–10 m s^{-1}) = –10 000 kg m s^{-1}
Before the collision the total momentum = 20 000 kg m s^{-1}

After the collision the mass of the vehicles is 2500 kg and their velocity is v m s^{-1}

The momentum after the collision = 2500 × v

If we can ignore any external forces the momentum before and after the collision must be the same:

20 000 = 2500 × v

$$v = \frac{20\,000}{2500} = 8 \text{ m s}^{-1}$$

> **E** Remember that momentum is a vector quantity. If the momentum in one direction is positive, the momentum in the opposite direction must be negative.

23

> **E** Linear momentum is conserved in both elastic and inelastic collisions, provided there is no resultant external force.

Collisions are classified as **elastic** or **inelastic**.

> **D** *In an elastic collision there is no loss of kinetic energy.*

If a collision is elastic, the total kinetic energy is the same before and after the collision. In an inelastic collision, kinetic energy is transferred to energy in different forms, such as heat. All collisions between everyday objects are inelastic – some energy is always transferred to other forms. Elastic collisions can take place between the molecules in a gas or between subatomic particles.

Elastic and inelastic collisions can be investigated in a school laboratory using gliders which run on an air track. The track has air blown through it so that the gliders rest on a cushion of air, eliminating friction between the glider and the track. If the air track is carefully levelled so that the glider moves horizontally, gravity will have no effect on the motion and we can say that there are no significant external forces acting on the glider. Light gates are used to time the glider's motion and a datalogger is often used to calculate and store the velocity of the gliders so that the motion can be analysed (Fig 27).

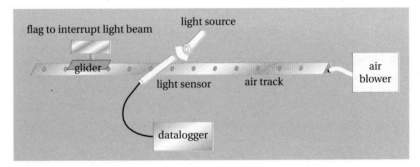

Fig 27 Experimental apparatus to demonstrate elastic collisions

Rockets

The operation of a rocket motor can be explained by the conservation of momentum. The rocket motor propels hot gases out of the nozzle. The rocket must gain an equal momentum in the opposite direction (Fig 28).

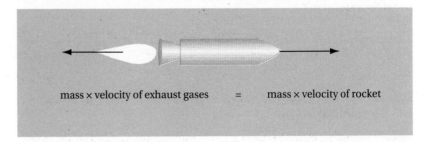

Fig 28 Conservation of momentum in a rocket

Example

A space-craft of mass 20 000 kg which is at rest fires its rockets. The exhaust gases are expelled at a rate of 100 kg s^{-1} and at a speed of 1000 m s^{-1}. If the rocket is fired for 10 seconds, what will the final velocity of the spacecraft be?

Answer

Total momentum before firing is zero. Therefore total momentum afterwards is also zero, the 'reverse' momentum of the exhaust gases must be balanced by the forward momentum of the rocket.

In 10 s the mass of exhaust gases is 100 kg s^{-1} × 10 s = 1000 kg.

Total momentum of gases = mv = 1000 kg × 1000 m s^{-1} = 1 000 000 kg m s^{-1}

Momentum of spacecraft must also be 1 000 000 = 20 000 × v, so v = 50 m s^{-1}

It is a common misconception that rockets work by pushing against the Earth. This can't be true, because rockets work out in space where there is no external object to push against. Rockets actually push against the exhaust gases, which push back with an equal but opposite force on the rocket.

This is dealt with in greater detail in 11.1.8, Newton's third law.

11.1.8 *Newton's laws of motion*

Newton's first law

Every object will continue to move with uniform velocity unless it is acted upon by a resultant external force.

This law expresses the idea that objects will stay at rest, or keep moving in a straight line at a steady speed, unless an external force acts on them. The law restates Galileo's law of *inertia*. Inertia is the reluctance of an object at rest to start moving, and its tendency to keep moving once it has started. This law isn't immediately apparent on Earth. If you give an object a push, it doesn't keep going in a straight line for ever, because on Earth it is difficult to avoid external forces like gravity or friction. In space, well away from any gravitational attractions, objects just keep moving in a straight line.

Another way of stating Newton's first law is to say that an object will remain in equilibrium, unless it is acted upon by an external force.

Newton's second law

The rate of change of an object's linear momentum is directly proportional to the resultant external force. The change in momentum takes place in the direction of the force.

Newton's second law defines a force as something that changes an object's momentum. Newton's second law can written:

Force, F, is proportional to the change in momentum divided by the time taken for the change.

$$F \propto \frac{\Delta p}{\Delta t} \quad \text{or} \quad F \propto \frac{\Delta(mv)}{\Delta t}$$

Newton's laws of motion were published in his *Principia* in 1687. His work on mechanics built on that of Galileo.

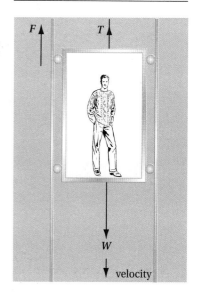

Fig 29
An example of Newton's first law
If the lift is to keep moving at a steady velocity, its weight must be balanced by the friction, F, and the tension, T. For equilibrium, $W = F + T$

This can be written $F = \dfrac{k\Delta(mv)}{t}$, where k is a constant.

If the mass of an object does not change then m is constant and:

$$F = \dfrac{km\Delta v}{\Delta t}$$

but $\dfrac{\Delta v}{\Delta t}$ = acceleration, a, so $F = kma$

The unit of force, the newton (N), is defined to be equal to 1, when $m = 1\,\text{kg}$ and $a = 1\,\text{m s}^{-2}$. This means that $k = 1$ and we can write:

$$F = ma$$

force (N) = mass (kg) × acceleration (m s^{-2}).

$F = ma$ for a constant mass

> **D** The S.I. unit of force is the newton, N. One newton is the force that will accelerate a mass of 1 kg by 1 m s^{-2}.

This way of writing Newton's second law allows us to calculate the effect of a force on an object.

Example

The total mass of a lift and its passengers is 1000 kg. The tension in the cable pulling the lift up is 12 000 N. Find the acceleration of the lift. (Take g = 10 N kg^{-1})

Answer

Upward force = 12 000 N

Downward force = mg = 1000 kg × 10 N kg^{-1} = 10 000 N

Resultant upwards force = 2000 N

Using $F = ma$, $a = \dfrac{2000}{1000} = 2\,\text{m s}^{-2}$

E Remember to find the resultant force first, then use Newton's second law to find the acceleration.

Newton's second law also explains why it is necessary for a car to decelerate relatively slowly, since stopping in a short time can lead to high forces. Seat belts, air-bags and crumple zones in cars are all ways of increasing the time taken to come to a stop during a crash. This reduces the force.

Water jets

In some cases, such as rockets and jets, mass cannot be treated as constant. It is then useful to think of force as the rate of change of momentum, and to apply Newton's second law in the form:

$$F = \dfrac{\Delta(mv)}{\Delta t}$$

Example

A pressure washer can eject 15 litres of water per minute at a speed of 50 m s^{-1}. The water jet hits a wall and comes to rest. Find the force exerted by the jet on the wall. (Density of water = 1000 kg m^{-3})

Answer

In 1 second, the pressure washer ejects $\frac{15}{60}$ = 0.25 litres of water.

Since 1 litre is 1×10^{-3} of a cubic metre, this is a mass of
$0.25 \times 10^{-3} \times 1000 = 0.25$ kg s^{-1}

The water is moving at 50 m s^{-1} so the momentum lost by the water in 1 second is:

$p = mv = 0.25 \times 50 = 12.5$ kg m s^{-1}

Since force in newtons is equal to the change of momentum in 1 second, the force is 12.5 N.

Newton's third law

If an object, A, exerts a force on a second object, B, then B exerts an equal but opposite force back on object A.

This law states that force between two bodies always acts equally on both objects, though in opposite directions. When we say that someone weighs 500 N, we mean that the gravitational attraction of the Earth on the person is 500 N. The person also attracts the Earth upwards with a force of 500 N. When a car exerts a force on the ground through the friction between its tyres and the road surface, the car pushes the ground backwards, whilst the ground pushes the car forwards. The two forces involved in Newton's third law never cancel each other out, because they act on different bodies.

This law is sometimes written as 'every action has an equal but opposite reaction', but you need to be careful with this statement. Remember that the forces act on different bodies.

The pair of forces under Newton's third law are always both of the same type, e.g. they may both be gravitational.

Example

A man stands still on the surface of the Earth. Draw the forces acting on the man and on the Earth and explain which forces are equal to each other.

Answer

$W_p = W_E$ By Newton's third law, the weight of the person is equal to the gravitational attraction of the person acting on the Earth.

$R_p = R_E$ By Newton's third law, the contact force of the Earth pushing on the person is equal to the contact force of the person pushing on the Earth.

$W_p = R_p$ By Newton's first law, because the person is in equilibrium, the person's weight is balanced by the contact force of the Earth.

$W_E = R_E$ By Newton's first law, because the Earth is in equilibrium, the gravitational attraction of the person is balanced by the contact force of the person on the Earth.

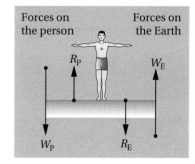

11.1.9 Work, energy and power

Work

Work is done whenever a force moves through a distance in the direction of the force. A force that does not move does no work. Sometimes the force is not in the same direction as the force, e.g. when a toy tractor is being pulled along by a string.

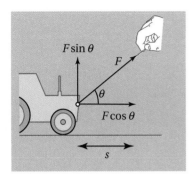

The force, F, can be resolved into two components: parallel to the motion and perpendicular to it. There is no movement in the direction of the perpendicular component, $F \sin \theta$, and so it does no work. All the work done is by the component that is parallel to the displacement, $Fs \cos \theta$.

> work done, W (J) = F (N) $\times$ s (m) $\times \cos \theta$

Work is measured in joules. One joule of work is done whenever a force of 1 newton moves through 1 metre.

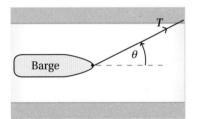

> *Example*
> *A barge is pulled along a canal by a tow rope. The tension in the tow-rope is 1000 N and it makes an angle of 20° with the canal. Find the work done in towing the barge 100 m.*
>
> *Answer*
> work done = $1000 \times 100 \times \cos 20° = 94$ kJ (to 2 s.f)

Energy and Power

> Energy is the ability to do work.

Energy can appear in many different forms. It may be classified as electrical, chemical, kinetic (movement), gravitational potential, thermal, radiant (e.g. light), nuclear or sound. Each of these forms of energy can be transferred so that the final effect is doing some work, such as the lifting of a weight. For example, electrical energy can be transferred as kinetic energy by an electric motor. This kinetic energy can be transferred as potential energy as the axle of the motor winds up a string and lifts a weight. At every stage some energy is transferred as random thermal energy, or heat. In this way, energy tends to become less concentrated and less useful. All forms of energy are measured in the same unit, the **joule**.

The rate at which a device can transfer energy is known as its **power**. Power is measured as the number of joules per second that are transferred. The S.I. unit of power is the **watt**.

> Power is the rate at which energy is transferred. A power of 1 watt means that 1 joule of energy is transferred every second.

A powerful electric motor can transfer thousands of joules of electrical energy into kinetic energy per second and its power would be given in kilowatts (kW).

> *Example*
>
> *An electric shower has an electrical power of 7 kW. If the shower runs for 10 minutes, how many joules of energy will have been transferred?*
>
> *Answer*
>
> The shower transfers electrical energy into thermal energy in the water at a rate of 7 000 joules per second. In 10 minutes, this is $10 \times 60 \times 7000 = 4\,200\,000$ J or 4.2 MJ.

Since energy can result in work being done, power can also be thought of as the rate at which work is done.

> **D** Power, P, is the rate at which work is done, $P = \dfrac{\Delta W}{\Delta t}$

> *Example*
>
> *A person of mass 60 kg runs up a flight of stairs that is 10 m high in 6 seconds, find the power output of the person. (Take $g = 10\,N\,kg^{-1}$)*
>
> *Answer*
>
> The person has to do work against the force of gravity to lift their own weight.
>
> force = $mg = 60 \times 10 = 600$ N
>
> The distance moved in the direction of the force is 10 m. Work done:
>
> $\Delta W = 600\,N \times 10\,m = 6000$ J
>
> If this work is done in 6 seconds, $\Delta t = 6$ s, $P = \dfrac{6000\,J}{6\,s} = 1000$ W or 1 kW.

For a moving machine, such as a motor or a car, it is often useful to relate the power output to the velocity at which the machine is moving. We can write:

$$\text{power} = \dfrac{\text{work done}}{\text{time}} = \dfrac{\text{force} \times \text{distance}}{\text{time}}$$

This can be written as power = force × $\dfrac{\text{distance}}{\text{time}}$

Since $\dfrac{\text{distance}}{\text{time}}$ = velocity, power = force × velocity, **$P = Fv$**

E This is the output power. The person would need a greater rate of energy **input**, since some energy would be transferred as heat. The person is not 100% efficient at transferring chemical energy into work.

E The force, F, and the velocity, v, must be in the same direction. If there was an angle of θ between the direction of the force and the velocity, the equation would become $p = Fv \cos \theta$.

Since Einstein's work on relativity, the conservation of energy has had to be extended to include mass. In module 1 you will have seen that energy can be transferred to mass in the process called pair production, where two particles are created from energy. Mass can also be transferred to energy in the process of annihilation when matter and anti-matter meet and are converted to gamma radiation.

11.1.10 Conservation of energy

When energy is transferred from one form to another, the total amount of energy does not change. This idea, known as the conservation of energy, is a fundamental principle in physics. It is not always obvious that the principle is obeyed. When a car brakes to halt at a junction it may appear that all the kinetic energy has disappeared, whereas in fact all the energy has been transferred to the brakes and the surroundings as heat. The conservation of energy only applies to a **closed system**, the total energy will not stay constant if energy has been transferred to another object. In the example of the braking car we have to take into account the energy transferred in heating the air and the road.

> The principle of conservation of energy states that the total energy of a closed system is constant.

Kinetic energy

The energy that a moving mass has because of its motion is known as its **kinetic energy**. The kinetic energy of a moving mass depends on the mass and on the velocity squared.

You will need to remember this equation.

> Kinetic energy (joules), $E_k = \frac{1}{2}mv^2$

The kinetic energy depends on velocity **squared,** rather than just velocity. So if a car doubles its speed, its kinetic energy will go up by a factor of four. This is why the stopping distance for a car goes up from 6 m at 20 mph to 24 m at 40 mph.

A common mistake is to square the whole expression rather than just the velocity, e.g.

$\left(\frac{1}{2} \times 1200 \times 8.9\right)^2$

which would give 2.85 MJ, far too big an answer. Make sure that you square the velocity first.

> *Example*
>
> *Estimate the average braking force needed if a family car is to achieve the braking distances above.*
>
> *Answer*
>
> The mass of a typical family car is just over a tonne, say 1200 kg. Since there are 1.6 km in a mile, a speed of 20 mph is
> $\frac{1600 \times 20}{60 \times 6} = 8.9 \, \text{m s}^{-1}$.
>
> The kinetic energy would be:
>
> $E_k = \frac{1}{2}mv^2 = 0.5 \times 1200 \times (8.9)^2 = 47\,400 \, \text{J}$
>
> This energy is used to do work against an average braking force F,
>
> so $E_k = F \times d$ or
>
> $F = \frac{E_k}{d} = \frac{47\,400}{6} = 7900 \, \text{N}$

Energy in collisions

When two objects collide the total momentum is always conserved. For some collisions, known as elastic collisions, the total kinetic energy is also conserved. Collisions between atoms and collisions between sub-atomic particles can be elastic. Collisions between macroscopic objects are never perfectly elastic, some of the kinetic energy is always transferred to thermal energy.

> *Example*
>
> *A car of mass 1500 kg travelling at 20 m s^{-1} crashes into another car of mass 1000 kg travelling in the opposite direction at 10 m s^{-1}. After the collision the cars stick together and move off at a speed of 8 m s^{-1}. Is the collision elastic or inelastic?*
>
> *Answer*
>
> Total kinetic energy before the collision =
>
> $\frac{1}{2} \times 1500 \times 20^2 + \frac{1}{2} \times 1000 \times 10^2 = 350\,000$ J
>
> Total kinetic energy after the collision = $\frac{1}{2} \times 2500 \times 8^2 = 80\,000$ J
>
> The total kinetic energy of the system is not constant, so this is an **inelastic** collision.

Kinetic energy is not a vector quantity. It makes no difference that the cars are travelling in opposite directions, the total kinetic energy is still the sum of their individual kinetic energies. This is not the case for momentum which is a vector quantity. Because the cars are moving in opposite directions, the total momentum is the difference in the car's momentum.

Gravitational potential energy

Gravitational potential energy is the energy that an object has because of its position in a gravitational field. You need to do work to raise a mass to a greater height above the surface of the Earth, this work is stored as potential energy. If the mass is released again, the potential energy will be transferred to the mass as kinetic energy.

The work done in lifting a mass, m, through a height, Δh, is:

ΔW = force × distance = $(m \times g) \times \Delta h$

This is also equal to the potential energy gained, ΔE_p, by the mass.

> **The change in the potential energy is the mass × gravitational field strength × the change in height. $\Delta E_p = mg\Delta h$**

You need to learn this equation.

The equation is only strictly true for places where the gravitational field strength, g, is constant. Although g does decrease as the distance from the Earth's surface increases, the equation is reasonably accurate for small values of Δh.

When a mass falls from a height, its potential energy is transferred to kinetic energy. If we can ignore energy losses due to air resistance, then **all** the potential energy will end up as kinetic energy; $\Delta E_p = \Delta E_k$

> *Example*
>
> *A high diving board is 10 m above the water surface. Calculate the speed at which a diver hits the water.*
>
> *Answer*
>
> $\Delta E_p = \Delta E_k \qquad mg\Delta h = \frac{1}{2}mv^2$
>
> So $v^2 = 2g\Delta h$
>
> $v^2 = 2 \times 10\,\text{N kg}^{-1} \times 10\,\text{m} = 200\,\text{m}^2\,\text{s}^{-2}$
>
> $v = \sqrt{200} = 14.1\,\text{m s}^{-1}$

11.1.11 Calculations involving change of energy

Internal energy

When energy is transferred to an object as thermal energy, its internal energy increases. In general the increase in internal energy will cause an increase in temperature. The temperature change caused by a given amount of energy depends on the mass of the object and on the material it is made of. Some materials require a lot of energy to cause a small temperature rise, whereas other materials need less. This property of materials is known as the **specific heat capacity, c,** it is measured in Joules per kilogram per degree Kelvin, $\text{J kg}^{-1}\,\text{K}^{-1}$.

E A temperature change of 1 degree Kelvin, 1 K, is identical to a temperature change of 1 C°. The Kelvin and Celsius scales have equal increments, they just start at different points, 0 K = −273.15 °C or 0 °C = 273.15 K.

D *The specific heat capacity of a material is the energy needed to cause a temperature rise of 1 K in a mass of 1 kg.*

The **heat capacity** of an object, a saucepan for example, is the energy required to raise its temperature by 1 C°. It depends on its mass and on the material it is made from. 'Specific' means the value per unit mass. In S.I. units, a unit mass is one kilogram, so specific heat capacity refers to 1 kg of a given material. Values for given materials can be looked up in data tables.

We can use the specific heat capacity to calculate the energy, ΔQ, required to heat any mass, m, of the material by any temperature rise, ΔT.

$\Delta Q = mc\Delta T$

Material	Specific Heat Capacity ($\text{J kg}^{-1}\,\text{K}^{-1}$)
Air	993
Water	4190
Copper	385
Concrete	3350
Gold	135
Hydrogen	14 300

Table 2
Examples of specific heat capacities.

> *Example*
>
> *A 2.2 kW electric kettle is used to heat 1.5 litres of water. Assume that all the energy is transferred as heat in the water, ignore the energy needed to warm the kettle itself and any heat losses. How long will it take the kettle to bring the water, initially at 10 °C, to the boil?*
>
> *Answer*
>
> The total energy required is:
>
> $\Delta Q = mc\Delta T = 1.5 \times 4190 \times (100 - 10) = 565\,650\,\text{J}$
>
> The electric kettle transfers 2200 Joules every second.
>
> Time needed $= \dfrac{565\,650}{2200} = 257\,\text{s}$

The specific heat capacity of water is relatively high, this means that water is useful for transferring energy for example in water-cooled engines. The high specific heat capacity of water also means that the temperature of a large mass of water, like the sea, only changes slowly.

Latent heat

When energy is transferred to an object as heat it does not always lead to an increase in temperature. The energy transfer can lead to a change in state, such as when ice melts or water turns to steam. The energy needed to change the state of a substance is known as its **latent heat, *l*.** The energy required to melt a substance is known as the latent heat of **fusion.** The energy required to convert a liquid to a gas (boil) a substance is known as the latent heat of **vaporisation.**

> The specific latent heat of fusion of a substance is the energy required to change 1 kg of a solid into 1 kg of liquid, with no change in temperature.

> The specific latent heat of vaporisation of a substance is the energy required to change 1 kg of a liquid into 1 kg of gas, with no change in temperature.

The energy, ΔQ, needed to change the state of a substance of mass m, is therefore;

$\Delta Q = ml$.

Material	Vaporisation (kJ kg^{-1})	Fusion (kJ kg^{-1})
water	2260	334
oxygen	243	14
helium	25	5
mercury	290	11
iron	6339	276
lead	854	25

Table 3
Specific latent heat values

When a substance changes from a solid to a liquid, or from a liquid to a gas, the average separation of its particles increases. Energy is needed to move the particles further apart, as work is done against the attractive forces holding the solid or liquid together. However, during a change of state there is no increase in the average kinetic energy of the particles, and so there is no change in temperature. Usually there is an increase in volume as a solid melts, or a liquid boils. Energy is needed to do work against external pressure as the substance expands. The latent heat of a substance is therefore the sum of the energy needed to increase the potential energy of its particles and to do work against external pressure.

When water turns to steam at atmospheric pressure (1×10^5 Pa) about 7% of the energy supplied is needed to do work against atmospheric pressure, the rest is needed to increase the potential energy of its molecules as they move apart from each other.

Fig 30

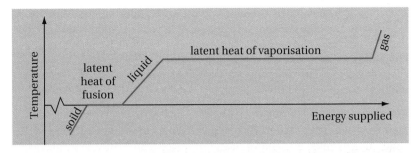

If energy is supplied at a constant rate to a solid substance its temperature will rise until it reaches its melting point. The substance will then melt at constant temperature. When all the solid has turned into liquid, the temperature will rise again until the liquid reaches its boiling point. The substance will then boil at constant temperature until all the liquid has turned into gas, when the temperature will rise again.

The energy needed to change the state of a substance is often used to dissipate thermal energy. The cooling towers of a power station transfer energy from the power station by evaporating large amounts of water. People do the same thing on a smaller scale, when we sweat we are transferring excess thermal energy by evaporating liquid.

Example

An ice cube of mass 20 g is added to 200 g of water which is initially at 20 °C. As the ice melts it has a cooling effect on the water. If all the energy needed to melt the ice comes from the water, calculate the cooling effect.

Answer

The latent heat necessary to melt the ice is
$\Delta Q = ml = 0.020 \text{ kg} \times 334 \times 10^3 \text{ J kg}^{-1} = 6680 \text{ J}$

If this energy all comes from the water:
$$\frac{\Delta Q}{mc} = \Delta T = \frac{6680 \text{ J}}{(0.200 \text{ kg} \times 4200 \text{ J kg}^{-1} \text{ K}^{-1})}$$

$\Delta T = 7.95 \, °C$

The temperature of the water will drop to about 12 °C because of the ice melting. The melted ice is now water at 0 °C. There will be another small cooling effect caused by the energy needed to raise 20 g of water at 0 °C to the final temperature of the drink.

Example

The amount of water that we lose through the evaporation of sweat depends on the temperature of our surroundings, as well as on the humidity and wind speed. On average a typical person loses about 0.5 litre of sweat per day. Calculate the average power of this energy transfer.

Answer

The energy needed to evaporate 0.5 l of sweat is $\Delta Q = ml$.

1 litre of water has a mass of 1 kg, so $m = 0.5$ kg.

l is the specific latent heat of vaporisation of water = 2.425×10^6 J kg^{-1}.

So the energy transferred, $\Delta Q = 0.5$ kg $\times 2.425 \times 10^6$ J kg^{-1} = 1.21×10^6 J.

If this is transferred over a period of 24 hours or $24 \times 60 \times 60 = 86\,400$ seconds, the average power is:

$$\frac{1.21 \times 10^6}{86\,400} = 14 \text{ W}.$$

Energy transfers

When energy is transferred from one form to another, for example from gravitational potential energy to kinetic energy, some of the energy inevitably ends up as random thermal energy in the surroundings. When a hammer is used to strike metal, potential energy is transferred to kinetic energy, but the end result is that the hammer and the metal get hotter. The brakes in a car transfer kinetic energy to thermal energy and the brakes can get red hot in the process.

The English scientist James Joule demonstrated that kinetic energy could be transferred to thermal energy by using horses to turn paddle wheels in a tank of water. A temperature rise of the water showed that energy had been transferred.

Example

A laboratory method of demonstrating this transfer uses lead shot in a closed cardboard tube. The tube is held vertically and turned end to end so that the lead is continually being lifted and allowed to fall, hitting the bottom of the tube. The tube is 1 m long and there is 250 g of shot in it. Lead has a specific heat capacity of 126 J kg^{-1} K^{-1}. How much of a temperature rise would you expect in the lead after 100 inversions of the tube?

Answer

Each time the lead falls, potential energy is transferred to kinetic energy and then to thermal energy. The change in potential energy = $mg\Delta h = 0.250$ kg $\times 9.8 \times 1$ m $= 2.45$ J, so 245 J after 100 inversions.

If this is all transferred to the lead, $\Delta T = \dfrac{\Delta Q}{mc}$

$$= \frac{245 \text{ J}}{(0.250 \text{ kg} \times 126 \text{ J kg}^{-1} \text{ K}^{-1})} = 7.8 \text{ °C}$$

In practice the temperature rise would be less than this because of heat losses to the surroundings.

11.2 Molecular kinetic theory

11.2.1 The equation of state for an ideal gas

The physical state of a fixed mass of gas can be described by three physical quantities: pressure, temperature and volume.

Pressure   The pressure, p, that a gas exerts on the walls of its container is caused by the collisions of the molecules with the walls. Pressure is defined as the force per unit area and is measured in pascals, Pa. 1 Pa is a pressure of 1 newton per square metre, $1\,\text{Pa} = 1\,\text{N}\,\text{m}^{-2}$.

Temperature   The temperature, T, of the gas is a measure of the average kinetic energy of its molecules.

Volume   The volume, V, is the space occupied by the gas, and is measured in m^3.

Temperature Scales

> E   The resistance of a metal wire, the wavelength of radiation emitted by a body and the volume of a gas are all properties which are used to measure temperature.

The temperature of a gas is measured by the change in the property of some other substance. This is known as a **thermometric property**. For example, a liquid in glass thermometer measures temperature in terms of the expansion of a length of liquid, usually mercury or alcohol. The thermometric property, length in the case of mercury thermometer, is measured at two **fixed points**. The change of state of a substance is often used as a fixed point in defining a temperature scale since the temperature is constant during a change of state. The Celsius temperature scale uses the change of state of pure water as its fixed points. 0 °C is defined as the melting point of pure ice and 100 °C is defined as the boiling point of pure water at a pressure of 1 atmosphere. The thermometric property, such as the length of a mercury thread, is assumed to vary linearly between these points and the scale is divided into 100 equal divisions.

Fig 31

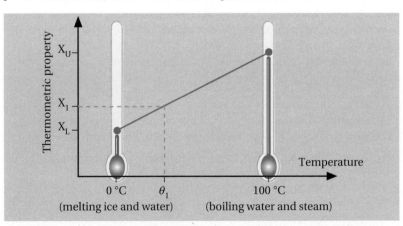

Unfortunately two different thermometers, perhaps one using mercury and one using alcohol, are only guaranteed to agree at the fixed points. This is because each thermometric property has its own temperature dependence; alcohol and mercury expand at different rates as they warm up. To overcome the problem of different temperature scales, a standard scale has been defined. This is the **absolute temperature** scale, also known as the **Kelvin** scale. The Kelvin scale is based on the behaviour of an **ideal** gas, a gas where there are no forces between its molecules.

As an ideal gas is cooled, its molecules slow down and the gas exerts a lower pressure on the walls of its container. If the container has a fixed volume, the pressure of the gas will continue to drop as the temperature drops.

> The forces between the molecules in a gas are relatively small because the molecules are much further apart than those of a liquid or a solid. A dry gas at low pressure behaves as an ideal gas.

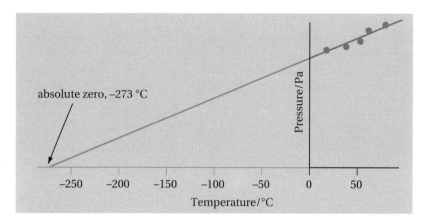

Fig 32
Fig 3 pg 218 Core

A graph plotted of temperature against pressure could be extrapolated backwards to very low temperatures. Eventually the pressure of an ideal gas would drop to zero. At this point the molecules have stopped moving and the gas cannot get any colder. This point is the lowest conceivable temperature and it is known as **absolute zero**.

The Kelvin scale uses the pressure of an ideal gas as its thermometric property. The two fixed points of the Kelvin scale are absolute zero and the triple point of water.

The triple point of water is defined to be 273.16 degrees on the Kelvin scale. This odd number is chosen to make a temperature change of one degree Celsius the same as one degree Kelvin. On the Kelvin scale absolute zero is 0 K (−273.15 °C) and the ice-point is 273.15 K (0 °C).

> It is theoretically impossible to cool something to absolute zero, but researchers have got very close. Liquid helium has been cooled to 90 μK (1 μK is one microkelvin or 10^{-6} K) at the University of Lancaster.

> The triple point of water is the unique combination of pressure and temperature when ice, water and water vapour all exist in thermal equilibrium. This only happens when the pressure is 0.6% of atmospheric pressure and the temperature is 0.01 °C.

> *Temperature in Kelvin = Temperature in Celsius + 273.15*

The Gas Laws

Most gases behave in similar ways. When the gas molecules are a long way apart, when the gas is at high temperature and low pressure, all gases behave as ideal and they are found to obey certain laws.

1. Boyle's law

When a gas is put under pressure, its volume decreases. For a fixed mass of an ideal gas at constant temperature, Boyle's law states that its volume is inversely proportional to its pressure.

> *For a fixed mass of an ideal gas at constant temperature* $p \propto \dfrac{1}{V}$

For a fixed mass of gas at constant temperature pV = constant, or $p_1V_1 = p_2V_2$

> *Example*
>
> *$20\,cm^3$ of air at atmospheric pressure, 1×10^5 Pa, is trapped in a bicycle pump when a finger is placed over the end of the pump. If the piston is pushed in until the volume is $5\,cm^3$, find the new pressure.*
>
> Answer
>
> $p_1V_1 = p_2V_2$ $\qquad 20\,cm^3 \times 1 \times 10^5\,Pa = 5\,cm^3 \times p_2$
>
> $p_2 = 4 \times 10^5$ Pa

Fig 33 Isotherms for an ideal gas

E Boyle's law only applies to a fixed mass of gas at constant temperature. You couldn't apply it to a car tyre or balloon that was being inflated since the new pressure and volume are affected by the increased number of gas molecules.

Each of these lines (isotherms) represents the behaviour of a gas at one specific temperature. At low temperatures, real gases can liquefy under pressure.

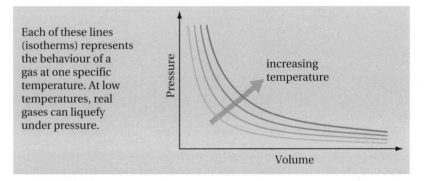

2. Charles' law

When gases are heated at constant pressure, they all expand at the same rate. The volume of a gas is proportional to its temperature, providing we use the Kelvin temperature scale.

> D *The volume of a fixed mass of an ideal gas at constant pressure is proportional to its absolute temperature, $V \propto T$.*

3. The pressure–temperature law

If a gas is heated in a container of fixed volume its pressure will increase. In fact the pressure of a gas is proportional to its absolute temperature.

> D *The pressure of a fixed mass of an ideal gas at constant volume is proportional to its absolute temperature, $P \propto T$.*

The equation of state

For a fixed mass of gas, the three gas laws can be summarised as:

$p \propto T$ $\qquad$ at constant volume

$V \propto T$ $\qquad$ at constant pressure

$p \propto \dfrac{1}{V}$ $\qquad$ at constant temperature

It is possible to combine these three laws into an equation of state for an ideal gas:

$pV \propto T$ or, for 1 mole of ideal gas, $pV = RT$

The constant R is known as **the molar gas constant**. $R = 8.314$ J mol^{-1} K^{-1}.

For **n** moles of gas the equation becomes: $pV = nRT$

> **D** The equation of state for an ideal gas is $pV = nRT$

E A mole is the SI unit of amount of substance. One mole of a gas contains 6.022×10^{23} molecules.

In many problems the mass of gas is fixed, such as in the cylinder of an engine or in a sealed balloon or car tyre. The initial values of the gas's pressure, volume temperature can be written, p_1, V_1 and T_1 and the equation of state can be written,

$$\frac{p_1 V_1}{T_1} = nR.$$

If the gas is compressed or heated then its pressure, volume and temperature will change to new values, p_2, V_2 and T_2, but n remains fixed. The equation of state is now:

$$\frac{p_2 V_2}{T_2} = nR.$$

This gives a useful form of the equation,

$$\frac{p_1 V_1}{T_1} = \frac{p_2 V_2}{T_2}$$

> *Example*
>
> *A bubble of air escapes from a diver's breathing apparatus at a depth of 45 m. The bubble has a volume of 2.0×10^{-5} m^3. The water pressure at a depth of 45 m is 450 kPa and the water temperature is 5 °C. What is the volume of the bubble when it has risen to the surface, where the temperature is 10 °C? Take atmospheric pressure as 100 kPa.*
>
> Answer
>
> The total pressure at a depth of 45 m is 550 kPa, 450 kPa due to the water and 100 kPa due to atmospheric pressure.
>
> The initial conditions of the gas are: $p_1 = 550$ kPa ; $V_1 = 2.0 \times 10^{-5}$ m^3; $T_1 = 278$ K
>
> The final conditions of the gas are $p_2 = 450$ kPa ; V_2 is unknown and $T_2 = 283$ K
>
> The mass of air in the bubble is fixed so $\frac{p_1 V_1}{T_1} = \frac{p_2 V_2}{T_2}$
>
> $\frac{p_1 V_1 T_2}{p_2 T_1} = V_2 = \frac{550 \text{ kPa} \times 2.0 \times 10^{-5} \text{ m}^3 \times 283 \text{ K}}{450 \text{ kPa} \times 278 \text{ K}}$
>
> $V_2 = 2.5 \times 10^{-5}$ m^3

E Don't forget that you must use the absolute temperature, i.e. the temperature must be in degrees Kelvin, not Celsius. Add 273 to the Celsius value to convert to Kelvin.

11.2.2 The molar gas constant R; the Avogadro constant, N_A.

Avogadro's law states that, at the same temperature and pressure, equal volumes of gases contain equal numbers of molecules. This is in agreement with the equation of state for an ideal gas.

Since $pV = nRT$, then $n = \dfrac{pV}{RT}$.

If the pressure, volume and temperature are the same for any two gases then **n**, the number of moles, must also be the same. Since a mole of substance always contains the same number of particles, Avogadro's law and the ideal gas equation are equivalent alternatives.

A mole is the SI unit of amount of substance. One mole of something contains as many particles (these could be atoms, ions, electrons or molecules) as there are atoms in 12 g of the carbon isotope $^{12}_{6}C$. The number of molecules in a mole of gas is always 6.022×10^{23}. This number is known as the Avogadro number, and is written N_A.

> The Avogadro number, N_A, is the number of particles in a mole of substance.
>
> $N_A = 6.022 \times 10^{23}$

The mass of 1 mole of a substance is its relative molecular mass expressed in grams. A mole of hydrogen therefore has a mass of only 1 g whilst a mole of carbon–12 has a mass of 12 g.

Example

A car tyre has a volume of around 1.50×10^{-2} m³ and it is inflated to a pressure that is twice atmospheric pressure. If the temperature is 20 °C, estimate the mass of air in the tyre. (The relative molecular mass of air is about 29.)

Answer
$$\dfrac{pV}{RT} = n = \dfrac{200 \times 10^3 \text{ Pa} \times 1.50 \times 10^{-2} \text{ m}^3}{8.314 \text{ mol}^{-1} \text{ K}^{-1} \times 293 \text{ K}} = 1.23 \text{ mole}$$

A mole of air would have a mass of 29 g so the mass of air in the tyre is about $1.23 \times 29 = 36$ g.

The average kinetic energy of molecules in a gas

As the molecules of a gas move and collide the total kinetic energy is shared amongst the molecules. The velocity of any particular molecule in a gas is random, it may be in any direction and it may have any one of a range of values. However the average kinetic energy of the molecules is proportional to the temperature of the gas. Kinetic energy is calculated as $\tfrac{1}{2}mv^2$, so if the average squared speed of a gas molecule is written $\overline{c^2}$ (the bar denotes the average value and the expression is pronounced 'c-squared bar') we can write:

$$T \propto \dfrac{1}{2}m\overline{c^2}$$

This suggests that at absolute zero the molecules have no kinetic energy. In other words they have stopped moving. In fact this is never quite the case, quantum effects known as zero-point energy mean that the kinetic energy never falls entirely to zero.

11.2.3 Pressure of an ideal gas

An ideal gas is one which obeys the equation of state, $pV = nRT$, exactly. There are no gases which are exactly 'ideal', but real gases at low pressures and at temperatures well above their liquefying temperature come quite close to ideal behaviour. It is possible to imagine a theoretical gas that would obey the gas laws exactly. This theoretical gas has to have the following properties:

- Its molecules are very small. We can ignore the space taken up by the molecules in comparison to the space between molecules. In other words the total volume occupied by the molecules is very much smaller than the volume occupied by the gas.
- Its molecules have elastic collisions. There is no loss of kinetic energy when a gas molecule collides with the walls of its container.
- The collisions take very little time, compared to the time between collisions.
- The molecules do not exert any force on each other, except during collisions. There are no intermolecular forces of attraction.
- There are a large number of molecules, so that a statistical approach can be used.
- The molecules move in random directions.

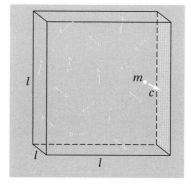

Fig 34

These assumptions are used to develop a kinetic theory of gases, which tries to predict the behaviour of a gas by considering the motion of its molecules. In particular we can derive an expression which links the pressure exerted by a gas with the speed of its molecules.

We start by imagining a single molecule, alone in a cubical box with sides of length l metres (Fig 34). The molecule has a mass m and is moving in a random direction with velocity c. The molecule will keep moving in a straight line, obeying Newton's first law of motion, until it has a collision with one of the walls of the container. It will exert a force on the wall before it bounces back and travels in the opposite direction (Fig 35). If we can find the force, and hence the pressure, exerted by one molecule, we can add up all the forces exerted by all the molecules in the box.

To simplify things a little further we will consider only the component of the velocity in the x-direction, this is perpendicular to the wall. The force exerted by a single collision can be found from Newton's second law of motion. The molecule has an initial momentum, mv_x, and after an elastic collision with wall the momentum will be the same size, but in the opposite direction, $-mv_x$.

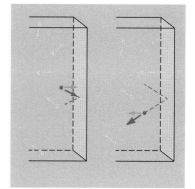

Fig 35

See page 25 for Newton's 2nd law.

The change in momentum = final momentum − initial momentum
$$= -mv_x - mv_x = -2mv_x$$

The molecule will hit this wall again after it has travelled to the other side of the box and back again, a distance of $2l$.

This will take a time of $\dfrac{2l}{v_x}$.

Force is defined as the rate of change of momentum, so the force, F, exerted by the molecule is:

$$F = \frac{\text{change in momentum}}{\text{time taken}} = \frac{-2mv_x}{2l/v_x} = \frac{mv_x^2}{l}$$

Pressure is force per unit area. The area of one wall is l^2, so $p = \dfrac{mv_x^2}{l^3}$

Since l^3 is the volume, V, of the box, this equation becomes: $p = \dfrac{mv_x^2}{V}$

The total pressure is the pressure from all of the other particles as well:

$$p = p_1 + p_2 + p_3 + \ldots = \frac{mv_{x1}^2}{V} + \frac{mv_{x2}^2}{V} + \frac{mv_{x3}^2}{V} + \ldots$$

$$p = \frac{m(v_{x1}^2 + v_{x2}^2 + v_{x3}^2 + \ldots)}{V}$$

The sum in the brackets is the total of all the molecules' squared velocities. This is equal to the average squared speed, $\overline{v_x^2}$, multiplied by the total number of molecules, N, in the box.

$$p = \frac{Nm\overline{v_x^2}}{V}$$
Equation 1

> **E** The mean velocity of all the molecules is always zero.

This is the pressure on a wall due to the x-component of the velocity of all the molecules in the box. Finally we must take into account the other components of velocity.

Fig 36

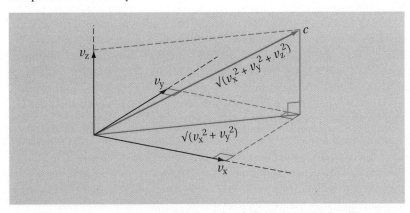

The speed of the particle c is linked to its components in the x, y and z directions by Pythagoras' theorem.

$$c^2 = v_x^2 + v_y^2 + v_z^2$$

Because the motion is random, the molecules are equally likely to be moving in any of the three directions so the average value of v_x^2, v_y^2 and v_z^2 will be the same: $\overline{c^2} = 3\overline{v_x^2}$

If we use this result in equation 1:

$$p = \frac{Nm\overline{c^2}}{3V} \quad \text{or} \quad pV = \frac{1}{3}Nm\overline{c^2}$$

This equation links the pressure and volume of a gas to the number of molecules and their average squared speed.

> **E**
> Since Nm is the total mass of the gas, $\frac{Nm}{V}$ is actually the density, ρ, of the gas. The equation can be written,
> $$p = \frac{1}{3}\rho\overline{c^2}$$

11.2.4 Internal energy; relation between temperature and molecular kinetic energy; the Boltzmann constant

Internal energy

The internal energy of a substance is the sum of the potential and kinetic energies of all its particles. In a gas the particles are so far apart that inter-molecular forces can be disregarded. The particles have no potential energy. The internal energy of a gas is therefore entirely due to the kinetic energy of the molecules. At any given time the total kinetic energy is shared randomly between all the molecules in the gas.

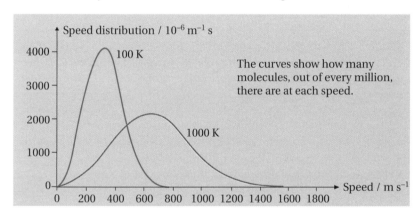

Fig 37

The curves show how many molecules, out of every million, there are at each speed.

When no energy is being transferred as heat between an object and its surroundings, it is said to be in **thermal equilibrium** with its surroundings. If a gas is in thermal equilibrium, and it is not being compressed or expanded, the average kinetic energy of its molecules will remain constant and the temperature will stay the same. Two objects at the same temperature are in thermal equilibrium.

>
> Two objects are in thermal equilibrium when there is no net flow of heat between them.

Thermometry relies on a law known as the Zeroth law of thermodynamics. It says that if two objects are in thermal equilibrium with a third object, they will be in thermal equilibrium with each other.

Temperature and molecular kinetic energy

The kinetic theory of gases gives a microscopic view of what is happening within a gas. It links the pressure and volume of a gas with the mass and speed of its molecules.

For 1 mole of gas kinetic theory leads to the equation:

$$pV = \frac{1}{3} N_a \overline{mc^2}$$

The equation of state for an ideal gas gives a macroscopic view, it links the pressure and volume of a gas to its temperature. For 1 mole of gas:

$$pV = RT$$

These two expressions should be equivalent and so:

$$\frac{1}{3} N_A \overline{mc^2} = RT$$

This equation lets us link the temperature of a gas to the average kinetic energy of its molecules, which is $½\overline{mc^2}$.

$$\frac{1}{2} \overline{mc^2} = \frac{3}{2}\left(\frac{R}{N_A}\right) T$$

> **E** The ratio $k = \dfrac{R}{N_A}$ is the molar gas constant divided by the number of atoms in a mole, so it could be regarded as the gas constant per molecule.

The ratio $\dfrac{R}{N_A}$ is called the Boltzmann constant, **k**. $k = 1.380 \times 10^{-23}$ J K^{-1}.

The equation $\dfrac{1}{2}\overline{mc^2} = \dfrac{3}{2}kT$

links the average kinetic energy of a gas molecule with the temperature of a gas.

Example

Calculate the average kinetic energy of a molecule of a gas at 20 °C. If the gas is oxygen, find the average squared velocity of a molecule. The relative molecular mass of oxygen is 32.

Answer

$$\frac{1}{2}\overline{mc^2} = \frac{3}{2}kT = 1.5 \times 1.380 \times 10^{-23} \text{ J K}^{-1} \times 293 \text{ K} = 6.0651 \times 10^{-21} \text{ J}$$

If the mass of 1 mole of oxygen is 32, then the mass of 1 molecule =

$$\frac{32}{N_A} = 5.31 \times 10^{-23} \text{ g}$$

or 5.31×10^{-26} kg. This gives a value for the mean squared velocity of:

$$\overline{c^2} = \frac{2 \times 6.0651 \times 10^{-21} \text{ J}}{5.31 \times 10^{-26}} = 2.28 \times 10^5 \text{ m}^2 \text{ s}^{-2}$$

> **E** The square root of this value is known as the root-mean-squared velocity, or rms velocity. In this example the r.m.s. velocity of an oxygen molecule is about 480 m s^{-1}

-- 2 Mechanics and Molecular Kinetic Theory

AS Sample module test

1   The diagram shows a straight, horizontal, uniform swimming bath spring board of length 4.00 m and of weight 300 N. It is freely hinged at A and rests on roller B, where AB is 1.60 m. A boy of weight 400 N stands at end C.

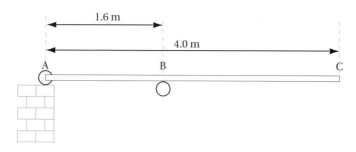

(a) (i) On the diagram above, show the direction of the forces acting on the board at A and B.
    (ii) Calculate the magnitude of the forces at A and B. (5)
(b) With the boy still on the board, a girl of weight 630 N also stands on the board. How far is she from A if the force at B is doubled? (2)

PH01 June 1996 Q1

2(a) (i) The diagram shows a view from above of two air-track gliders, A, of mass 0.30 kg, and B, of mass 0.20 kg, travelling towards each other at the speeds shown.

After they collide and separate, A travels with a speed if 0.20 m s^{-1} from right to left. Apply the principle of conservation of momentum to this collision, and hence find the velocity of B after the collision.

   (ii) What general condition must be satisfied for momentum to be conserved in a collision? (5)
(b) (i) Explain the difference between a *perfectly elastic collision* and an *inelastic collision*.
    (ii) Show that the collision described in part (a) is inelastic.
    (iii) Give **one** example of a perfectly elastic collision. (3)
(c) The speed of a bullet can be estimated by firing it horizontally into a block of wood suspended from a long string so that the bullet becomes embedded in the centre of the block. The block swings so that the centre of mass rises a vertical distance of 0.15 m. The mass of the bullet is 10 g and that of the block is 1.99 kg.
   (i) Assuming that air resistance can be neglected, calculate the speed of the block and bullet immediately after the impact.
   (ii) Calculate the speed of the bullet before the impact. (5)

PH01 June 1996 Q3

45

AQA (A) Physics AS

3(a) Distinguish between a scalar quantity and a vector quantity. (2)

(b) A car travels one complete lap around a circular track at an average speed of $100\,\text{km}\,\text{h}^{-1}$.

　i) If the lap takes 3.0 minutes, show that the length of the track is 5.0 km.

　ii) What is the total displacement of the car at the end of the lap? (3)

PH01 Feb 1996 Q1

4 Whilst braking, a car of mass 1200 kg decelerates along a level road at $8.0\,\text{m}\,\text{s}^{-2}$.

(a) If air resistance can be assumed to be negligible, calculate the horizontal force of the road on the car. (1)

(b) Calculate the vertical force exerted by the road on the car. (1)

(c) Using a scale drawing or otherwise, calculate to the magnitude and direction of the resultant force of the road on the car. (5)

PH01 Feb 1996 Q3

5(a) The table of results is taken from an experiment where the time taken for a ball to fall vertically through different distances is measured.

distance fallen/m	1.0	2.0	4.0	6.0	8.0	10.0
time/s	0.45	0.63	0.90	1.10	1.30	1.40

　(i) Plot a graph of distance fallen on the *y*-axis against time on the *x*-axis.

　(ii) Explain why your graph is not a straight line.

　(iii) Calculate the gradient of the graph at 0.70 s.

　(iv) State what your gradient represents and use it to determine a value for the acceleration of free fall, assuming that the ball started from rest. (9)

(b) A falling body experiences air resistance and this can result in it reaching a terminal speed. Explain, using Newton's law of motion, why a terminal speed is reached. (5)

PH01 Feb 1997 Q4

6 A dart is thrown at an angle of 20° above the horizontal and at a speed of $5.0\,\text{m}\,\text{s}^{-1}$. It strikes a dart board, situated a horizontal distance of 2.5 m from its point of projection.

(a) Calculate

　(i) the horizontal component of the velocity of the dart,

　(ii) the time taken for the dart to reach the board. (3)

(b) If the dart is travelling horizontally at the time it strikes the board, calculate the vertical displacement of the dart. (3)

PH01 Feb 1997 Q6

7(a) (i) Write down the equation of state for n moles of an ideal gas.

(ii) What is meant by an ideal gas? (3)

(b) Calculate the mass of argon gas filling an electric light bulb of volume $8.2 \times 10^{-5}\,\text{m}^3$ if the pressure inside the bulb is 90 kPa and the temperature of the gas is 340 K.
   density of argon at standard temperature and pressure = $1.56\,\text{kg m}^{-3}$
   standard pressure = 100 kPa
   standard temperature = 273 K (4)

PH03 June 1996 Q5

8(a) A single gas molecule of mass m is moving in a rectangular box with a velocity of u_x in the positive x-direction as shown in the diagram. The molecule moves backwards and forwards in the box, striking the end faces normally and making elastic collisions.

(i) Show that the time, t, between collisions with the shaded face is $t = \dfrac{2l_x}{u_x}$.

(ii) If it is assumed that the box contains N identical molecules, each of mass m, all moving parallel to the x-direction with speed u_x and making elastic collisions at the ends, show that the average force, F, on the shaded face is given by
$$F = \frac{Nmu_x^2}{l_x}$$

(iii) In a better model of molecular motion in gases, molecules of mean square speed $\overline{c^2}$ are assumed to move randomly in the box. By considering this random motion, show that a better expression for F is
$$F = \frac{Nm\overline{c^2}}{3l_x}$$
and hence derive the equation
$$pV = \frac{1}{3}Nm\overline{c^2}$$
(9)

(b) Take $R = 8.31\,\text{mol}^{-1}\,\text{K}^{-1}$ in this part if the question.

For 1.0 mol of helium ($M_r = 4.0$) at a temperature of 27 °C, calculate

(i) the total kinetic energy of the molecules,

(ii) the root mean square speed of the molecules. (4)

(c) 1.0 mol of the gas neon is mixed with the 1.0 mol of helium in part (b). Calculate the total kinetic energy of the molecules at 27 °C. (2)

PH03 June 1998 Q5

1(a)(i)

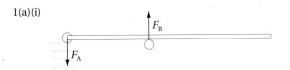

(ii) taking moments about A
(400 × 4.0) + (300 × 2) = F_B × 1.6   1
2200 = F_B × 1.6 so F_B = 1375 N   1
F_A = 1375 − 700 = 675 N   1
allow e.c.f. from F_B
No marks if weight on board not used

(b) taking moments about A
2 × 1375 × 1.6 = (630 × d) + (300 × 2) + (400 × 4)   1
allow e.c.f. from (a)(ii)
2200 = 630 × d so d = 3.5 m

2(a)(i) (0.30 × 0.20) − (0.20 × 0.50) = (0.30 × 0.20) + (0.20 × v_b)
                                               1 [if all correct]
correct signs   1
solution gives v_b = 0.10 m s^{-1} 1
left to right, stated or from diagram   1

(ii) no external force [closed system, no friction or other forces]   1

(b)(i) perfectly elastic − total k.e. unchanged by collision
inelastic − (some) k.e. changed to other forms   1

(ii) k.e. of A after = k.e. of A before (same speed)   1
speed of B after < speed of B before so loss in k.e. (inelastic)   1
alternative:
k.e. before = 0.031 J
relative speed before = 0.7 m s^{-1} 1
k.e. after = 0.007 J
relative speed after = 0.3 m s^{-1} which is less   1

(iii) gas molecule collisions
magnetic repulsion of gliders on air track
alpha particle scattering
neutron scattering by moderator   1

(c)(i) conservation of energy **after** collision
or ½mv^2 = mgh   1
hence v^2 = 2 × 9.8 × 0.15 [= 2 × 10 × 0.15]   1
hence v = 1.71 m s^{-1} [= 1.73 m s^{-1}] 1
(no marks for if constant acceleration equations used)

(ii) conservation of momentum
mV = (m + M)v [or correct substitution of values]   1
V = 343 m s^{-1} [= 346 m s^{-1} if g = 10]   1

3(a) scalars have **magnitude** (or **size**)   1
vectors have **magnitude and direction**   1

(b)(i) s = vt   1
s = 100 × 3/60 = 5 km   1

(ii) zero **km** (or other correct unit)   1

4(a) F = ma = 1200 × 8.0 = 9600 N   1
(b) F = ma = 1200 × 9.81 = 11800 N   1
(c) either by calculation:
$F_{res} = \sqrt{(F_1)^2 + (F_2)^2}$ 1
F_{res} = 5200 N   1
$\tan\theta = F_2/F_1$ 1
θ = 51°   1
with some indication of direction   1
(lose last two marks if 11800 N is shown acting down)

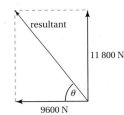

or by drawing
resultant = 15200(±200) N   1
θ = 51°   1
scale stated   1
some indication of direction   1

5(a)(i) suitable scales, labelled axis, five correctly plotted points, smooth continuous curve   1 each
(ii) because the ball is **accelerating**   1
(iii) tangent drawn at 0.70 s ± half a square   1
correct calculations   1
gradient = 7.3 (±0.2) m s^{-1} 1
(iv) velocity (at 0.7 s)
g = candidate's gradient divided by 0.70   1

(b) the force of gravity causes acceleration   1
increasing speed results in increasing air resistance   1
when air resistance equals weight   1
there is no resultant force   1
and therefore constant speed   1

6(a)(i) v = 8.7 cos 20°   1
v = 8.2 m s^{-1} 1
(ii) t = 2.5/8.2 so t = 0.30 s   1
(b) vertical component = 8.7 sin 20°   1
2 × 9.81 × s = (8.7 sin 20°)^2 1
s = 0.45 m   1

7(a)(i) pV = nRT
(ii) Boyle's law
obeys pV = nRT } 2
gas laws   } 1
at all temperatures and pressures   1
or molecules of negligible volume
with negligible forces between them } 1

(b) p/ρT = constant   1
100 × 10^3/1.56 × 273 = 90 × 10^3/ρ_2 × 340
gives ρ_2 = 1.13 (kg m^{-3})   1
mass of gas = $\rho_2 V$ = 1.13 × 8.2 × 10^{-5} =
9.2(7) × 10^{-5} kg   1
alternative calculations also give the correct answer

8(a)(i) molecule travels $2l_x$ at u_x, t = d/v = $2l_x/u_x$ 1
(ii) change of momentum per collision = Δp = $2mu_x$ 1
force = rate of change of momentum or $\Delta p/\Delta t$ 1
= $2mu_x \times 2l_x/u_x$ 1
for N molecules, force = ($\Delta p/\Delta t$)N = $2mu_x \times u_x/2l_x \times N$ 1
= Nmu_x^2/l_x
(iii) for typical molecule: $c^2 = u_x^2 + u_y^2 + u_z^2$
i.e. $\overline{c^2} = \overline{u_x^2} + \overline{u_y^2} + \overline{u_z^2}$ 1
$\overline{u_x^2} = \overline{u_y^2} = \overline{u_z^2}$ (for random motion)   1
so $\overline{u_x^2} = \overline{c^2}/3$ and F = ⅓($mN\overline{c^2}/l_x$)   1
p = F/A = ⅓($mN\overline{c^2}/l_xl_yl_z$)   1
and $l_xl_yl_z$ = V (hence result)   1

(b)(i) E_{mol} (= 3/2RT) = 3/2 × 9.31 × 300
= 3740 J [or 3.7 × 10^3 J]   1
(ii) $\overline{c^2}$ = 3RT/M   1
$\sqrt{(\overline{c^2})} = \sqrt{(3 × 8.31 × 300/4.0 × 10^{-3})}$ = 1.37 × 10^3 m s^{-1} 1
(c) adding 1 mol neon gives 2 mol gas, doubling E   1
= 7480 J [or 7.5 × 10^3 J]   1